Driving Test

GETTING IT RIGHT

Driving Test

Revised by
Chris Goffey

Photography by
Jeff Wright

Foreword by
Noel Edmunds

foulsham
LONDON • NEW YORK • TORONTO • SYDNEY

foulsham
Yeovil Road, Slough, Berkshire, SL1 4JH

ISBN 0-572-01778-2

Copyright © 1993 W. Foulsham & Co. Ltd.

All rights reserved.

The Copyright Act prohibits (subject to certain very limited exceptions) the making of copies of any copyright work or of a substantial part of such a work, including the making of copies by photocopying or similar process. Written permission to make a copy or copies must therefore normally be obtained from the publisher in advance. It is advisable also to consult the publisher if in any doubt as to the legality of any copying which is to be undertaken.

Typeset in Great Britain by Typesetting Solutions, Slough, Berks.
Printed in Great Britain by Cox and Wyman Ltd., Reading.

CONTENTS

Foreword		6
About This Book		7
Chapter 1	Are you fit to drive?	8
Chapter 2	At home	11
Chapter 3	Changing gear	16
Chapter 4	Stop! It's brake time	34
Chapter 5	Steering guidelines	41
Chapter 6	Look backwards forwards	48
Chapter 7	'Starting away' practice	51
Chapter 8	Turning right	55
Chapter 9	Turning left	60
Chapter 10	In reverse	65
Chapter 11	Getting ready	71
Chapter 12	On the move	78
Chapter 13	Observation	89
Chapter 14	Choosing an instructor	95
Chapter 15	How to pass the driving test	98
Chapter 16	Basic maintenance	104
Chapter 17	First aid	117
Rating test answers and rating test graph		126

FOREWORD

I've been driving for 25 years now. I really enjoy it — you may even have seen pictures of me driving in rallies and races. But if I had to learn to drive again, I'd be a bit worried about two things.

The roads are getting more and more crowded (which makes safe driving more important than it's ever been).

The rules and regulations are getting more and more complex.

Here at last is a book which lays the foundations for learning to be a really good driver in contemporary conditions.

What's more it shows you how to get fun out of driving — as much fun as I've had.

ABOUT THIS BOOK

Let's get one thing straight. No publication can teach you to drive. What this book will do is prepare you for on-the-road instruction so thoroughly, that you should need fewer driving lessons to reach test standard.

All you have to do is sit in a car — your own, your family's or a friend's. You don't start it or drive it. You will simply become practice perfect in your own drive with the basic exercises. In fact, don't have the key in the ignition.

In addition there are chapters on how a car works, basic maintenance, first aid, and advice on choosing an instructor. All through, too, there are rating tests to monitor your progress.

Three points: Always wear sensible, comfortable clothes and footwear for the exercises in this book and for driving (steer clear especially of shoes with chunky soles and heels). Get and study a copy of the DL 68 booklet, which sets out the requirements for the driving test and is obtainable from your driving instructor. Buy and learn straightaway the Highway Code (the law's demands at the back, too).

ARE YOU FIT TO DRIVE?

By the year 2000, there could be many more cases of cars failing to start and engines suddenly cutting out on the road. Not, it must be hastily explained, because of increased unpredictability due to falling manufacturing standards but because of . . . greater reliability.

Remarkably, vehicles could tell you that you aren't fit to drive . . . and ensure that you don't. Some production cars already have devices that prevent you from moving off until you have put on your seat-belt. That's nothing compared to development in the pipeline . . . particularly in America.

There are breath sensors to determine if you've had too much to drink, and pupillographs to assess alertness. Who knows? In years to come, we may also be obliged to connect electrodes to our chests to monitor heart condition and stress.

How much simpler and less expensive it would be if nations followed the Greeks' example of insisting on strict medical and eyesight tests before issuing driving licences. Surely, if their government considers it a necessary precaution and, what is more, haven't allowed their decision to be influenced by the threat of losing votes at election time, they must be convinced that ill-health and poor sight cause road accidents.

Britain, too, clearly recognises the dangers — drivers of buses and lorries must have sight and physical examinations. Also the hours they spend at the wheel are controlled by legislation. Yet apart from satisfying a government examiner at the driving test that he can read a vehicle number plate from 75ft (23m), a medical or peripheral vision check isn't required for the motorist.

The driving licence application form requires candidates to indicate whether they suffer from epilepsy, diabetes, fainting, blackouts, etc., pointing out that failure to admit the truth can lead to prosecution, but who's to know a false statement's been made until it is too late? Not only that; a person's state of health can change overnight. But, in the absence of official periodic checks how's this to be discovered?

Just how many road accidents are caused by disorders of health, sight and mentality isn't known. But there is evidence to suggest that some could be. The great majority of accidents are put down to driver error. It is known that more than 20% of industrial accidents are triggered by 'improper mental attitude or personality maladjustment'. In addition, 80% are influenced by 'general social factors', and it has been shown that the victims suffered 'tension-producing situations' immediately before the mishaps.

How does all this affect you? Simply by suggesting that, in the absence of legislation, you could be doing yourself a very big favour by asking your doctor and an optician whether there's any reason why you shouldn't

ARE YOU FIT TO DRIVE?

apply for a provisional driving licence. And, having been given the 'okay', to undergo similar checks every five years.

As the government itself points out: 'Health affects driving; even a cold can put you below par.'

On the subject of eyes, it adds: 'Vision changes, and we tend not to notice a gradual worsening in our sight. It is sensible to have a regular check-up.'

Try this simple test: stretch your arms out on either side at shoulder height with both thumbs up, and move your hands forward about six inches (15cm). If you cannot see both thumbs *while you are looking straight ahead,* then your side vision is restricted — and awareness of what is happening out of the corner of your eye is important to drivers.

There are other questions you should be asking yourself, too. Like: are you fit? Top-line physical co-ordination is vital for quick and correct reactions to driving situations. Do you have the right attitude? In other words, do you have a sense of responsibility for others' safety? Is your concentration, patience and courtesy all that it should be?

You may be in a high-powered job. If so, you could suffer a great deal of tension or emotional problems. Psychological tests have revealed that emotional or neurotic people tend to over-react to stress, decline in performance, and are accident-prone.

Even more alarming is another finding—that sufferers of mental fatigue go through 'unnoticed rest pauses', when, even though fully awake, they cannot pay attention to the task at hand. If that 'task at hand' is driving ... well, the dangers scarcely need spelling out.

Don't let it happen to you. Go along to your doctor and optician ... for everyone's peace of mind.

AT HOME

Getting used to the pedals

It's the reason why most learners at the start have what is known as 'kangaroo petrol' in their tank – those violent, embarrassing and downright off-putting lurches when moving off which, to add to the novice's misery, all too often end with the engine cutting out.

It's also one of the reasons why learners have difficulty in getting to grips with several manoeuvres they are required to carry out.

It's foot co-ordination ... or more accurately, the lack of it. But get it right from the off – and you will have cleared the biggest obstacle to smooth driving, and, most of all, you will have acquired confidence.

But if you are used to the way the pedals, the gear lever, and handbrake look and feel, and the correct sequence for operating them, before you start up and actually drive, you are really on your way. You'll pick up everything much faster in your lessons from a professional instructor.

What is foot co-ordination? In a nutshell, it's clutch and accelerator pedal control.

We won't complicate matters at this stage by going

AT HOME

into technical details of what happens under the bonnet when the appropriate pedal is depressed. All you need to know, for now, is that in a car with manual transmission there are three pedals operated by the feet. The clutch is the one on the left; the accelerator on the right.

(Between these is the most important foot control of the trio—the pedal that applies the brakes to all four wheels.)

We shall come to the brakes later . For starters, we will concentrate on the clutch and accelerator.

The clutch first. This is pushed down each time you change gear and is let up smoothly once the gear lever has been placed in the selected notch. The accelerator is simple, the further this is depressed, the quicker the car goes, while releasing the pedal will cause the road speed to decrease.

The main thing to remember is that you never push down both at the same time. Rather, as one pedal goes down, the other comes up. In other words: it's walking on the spot in a sitting position.

The pictures on the following three pages will give you the idea. Sit in the car, make sure you are comfortable, and that you can push the pedals to the floor without straining your legs too far forward. Slide the seat forward until you are able to depress the pedals without strain in either thigh or calf. Start by pushing down the left pedal, called the clutch. The one in the middle is the brake, and the one on the right the accelerator.

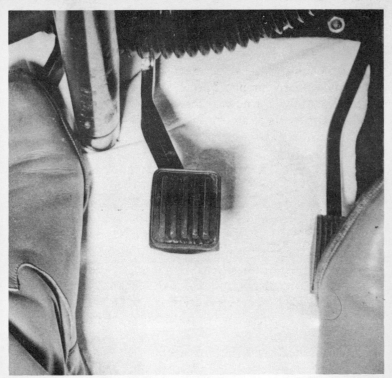

Start by pushing down the left-hand (clutch) pedal.

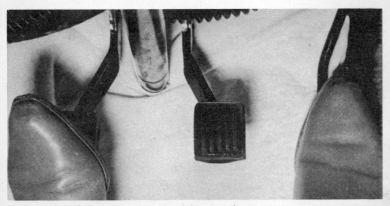

Then, as the left foot comes up, the right goes down.

As the right foot reaches the floor, the left is usually raised.

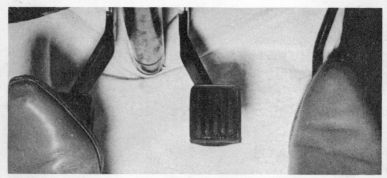

Continue the sequence by bringing the right foot up and, at the same time pushing the left down.

As the left foot reaches the floor, the right is fully raised.

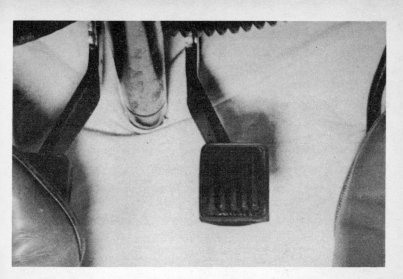

Now bring the left foot up, at the same time pushing the right down.

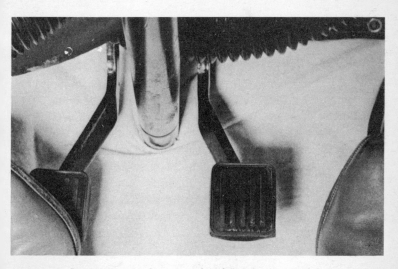

As the right foot reaches the floor, the left is fully raised ... and so on. Repeat over and over again.

CHAPTER 3

CHANGING GEAR

On, now, to what foot co-ordination is all about: changing gear.

This introduces the stick with a knob on the top sticking out of the floor on your left. It should have the positions of the gears marked on it. If it doesn't, look in the handbook, or get a friend to draw out a diagram for you. You may find the lever sticky, and difficult to move into gear with the engine off. Don't worry, it will be a lot easier when the car is moving; at this stage we are simply trying to give you the feel of the controls.

This exercise also introduces another aspect L-drivers find difficult—co-ordinating foot and hand movements. Keep practising the procedures on the next 13 pages until you can do them without looking at your feet or the gearbox. Each sequence goes from left to right.

Start with the gear in the central, neutral position (picture right, top); the left foot to the left of the clutch, the right behind the accelerator (picture right, bottom).

CHANGING GEAR

Fully depress the clutch with the left foot and, with the left hand grasp the gear lever which (far right) should be in the central, neutral position.

Keep the clutch down ... put the lever into the first gear position ... which is shown far right.

Take your hand off the gear lever. Now push the accelerator down a trifle and keep your right foot completely steady in that position, at the same time slowly bringing up the left foot until fully raised. First gear is now engaged (far right).

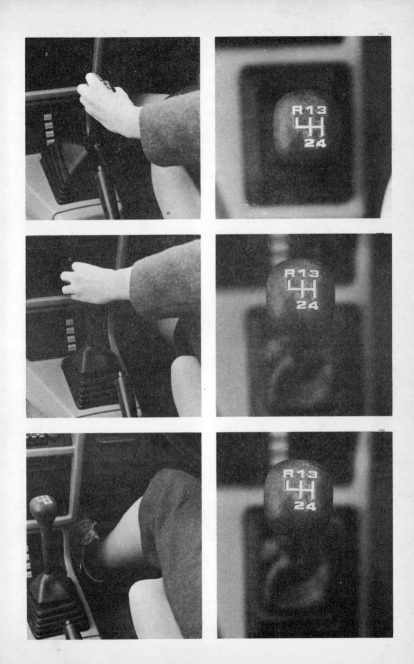

Slowly take your foot off the accelerator, then depress the clutch . . . grasp the lever . . . which (far right) is still in first gear.

Keep the clutch down . . . put the lever into second gear which is shown far right.

Take your hand off the lever and slowly bring up the clutch. At the half-way point start pushing down the accelerator, at the same time continuing to bring up the left foot until fully raised. Second gear is now engaged (far right).

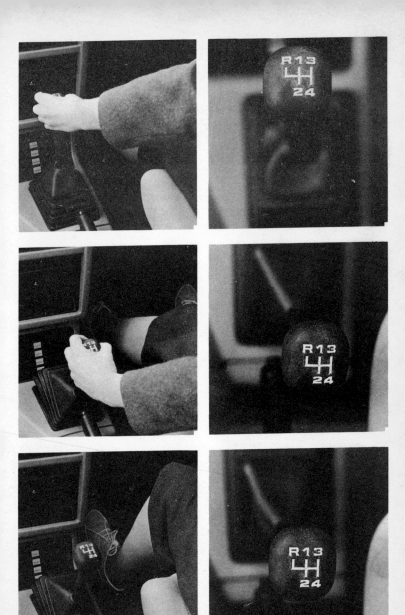

Take your foot off the accelerator, then depress the clutch ... grasp the lever ... which (far right) is still in second gear.

Keep the clutch depressed ... put the lever into the third gear position ... which is shown far right.

Take your hand off the lever and slowly bring up the clutch. At the half-way point, start pushing down the accelerator, at the same time continuing to bring up the left foot until fully raised. Third gear is now engaged (far right).

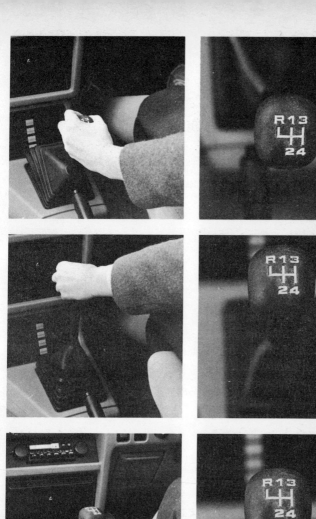
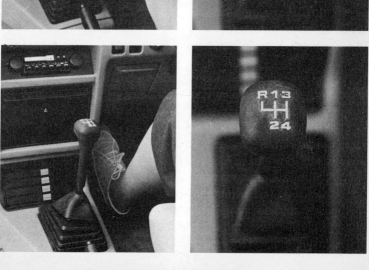

Ease your foot off the accelerator, then depress the clutch ... grasp the lever ... which (far right) is still in third gear.

Keep the clutch down ... put the lever into fourth gear ... which is shown far right.

Take your hand off the lever and slowly bring up the clutch. At the half-way point, start pushing down the accelerator, at the same time continuing to bring up the left foot until fully raised. Fourth gear is now engaged (far right).

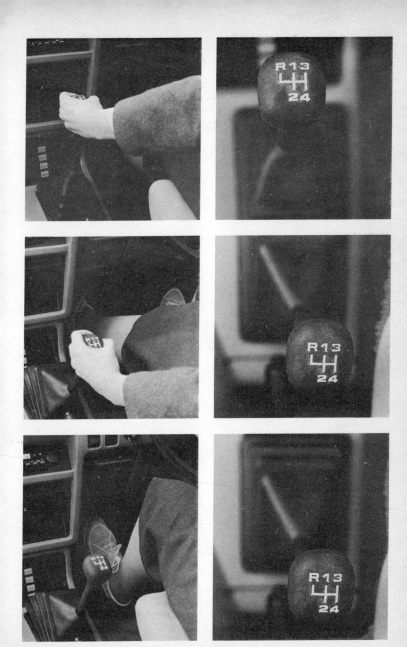

Ease your foot off the accelerator then depress the clutch . . . grasp the lever . . . which (far right) is still in fourth gear.

Keep the clutch down . . . put the lever into neutral . . . which is shown far right. Then take your foot off the clutch.

To obtain reverse gear, depress the clutch . . . grasp the lever . . . which (far right) is in neutral.

Keep the clutch down . . . put the lever into the reverse gear position . . . which is shown far right.

Take your hand off the lever. Now push the accelerator down a trifle and keep your right foot completely steady in that position, at the same time slowly bringing up the left foot until fully raised. Reverse gear is now engaged (far right).

Ease your foot off the accelerator, then depress the clutch . . . put the lever back into the neutral position . . which is shown far right. Then take your hand off the lever and your foot off the clutch.

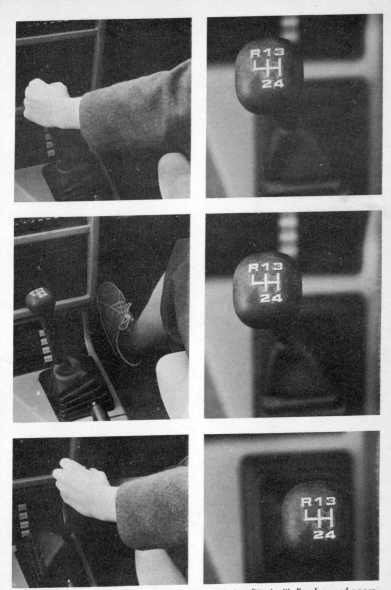

Note: for economy reasons, more cars now are being fitted with five forward gears. This should present no problems if you encounter such a car; you just need to make one further change, as described, from fourth into fifth gear.

CHANGING GEAR

Keep on practising...

Repeat the gearchanging exercise over and over (back through the gears from fourth to first, too).

Going back the other way is a similar procedure. If changing from fourth to third, it's foot off the accelerator, clutch down, gear lever into third, hand off lever, bring up the clutch and, at the half-way position, start pushing down the accelerator — at the same time continuing to bring up the left foot until fully raised.

Get into the habit of resting the left foot on the floor when the clutch is not in use (see below). The slightest pressure on the pedal while the car is being driven can result in premature clutch replacement — a costly business.

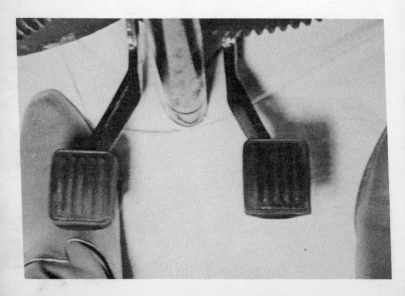

Rating Test: 1

(answers on page 126)

Knowing the Highway Code backwards and putting everything it preaches into practice at *all* times is the basic hallmark of good driving. This paper tests your knowledge of one of its sections — everyday traffic signs and road markings. Tick the appropriate definition in each case.

1a New continental level crossing with barrier.
 b Quayside ahead.
 c Opening or swing bridge.

2a Level crossing with other barrier or gate ahead.
 b Pedestrians use gated entrance.
 c Advance warning of cattle grid.

RATING TEST: 1

3a Pass either side.
 b Keep left unless you are turning right.
 c Motorway sign: Centre lane closed.

4a Traffic merging.
 b Change to opposite carriageway.
 c Special road works: Beware of deviations.

5a Warning of Give Way.
 b Advance warning of stop sign.
 c Temporary island for pedestrians.

6a Dual carriageway ends.
 b Road narrows on both sides.
 c Narrow bridge ahead.

RATING TEST: 1

7a Road Narrows.
 b Road narrows on offside.
 c Leaving dual carriageway.

8a Dangerous junction ahead.
 b Two-way traffic straight ahead.
 c Two-way traffic crosses one-way road.

9a No waiting.
 b No overtaking.
 c Outside lane closed.

10a Two-way traffic straight ahead.
 b Two-way traffic crossing one-way road.
 c Give priority to vehicles from opposite direction.

Your score out of 10 ....
(Mark the result on the graph on Page 128)

CHAPTER 4
STOP! IT'S BRAKE TIME

So far, we have explained about getting the car moving. Much more important from the safety viewpoint is being able to stop it.

A vehicle has two braking systems — a footbrake which applies the brakes to all four roadwheels, and a handbrake that puts on the rear brakes only.

As mentioned earlier, the footbrake is the middle pedal. And it is *always* operated by the right foot.

The handbrake, in contrast, is applied by a lever — usually mounted between the two front seats.

Have a look at the handbrake lever in your car — but make sure you are not on a gradient before you touch it or release it. If you have any doubt, put a brick or a block of wood against a wheel to stop the car rolling.

STOP! IT'S BRAKE TIME

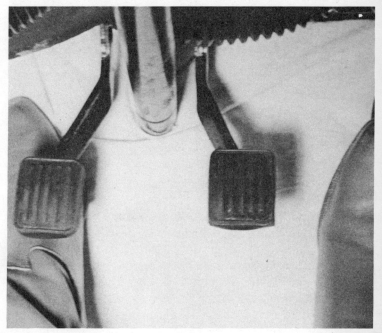

Begin this exercise as if you're driving. You are in fourth gear, your right foot is on the accelerator, and the left is 'resting'.

You see a pedestrian crossing the road ahead, so it's foot off the accelerator and cover the footbrake.

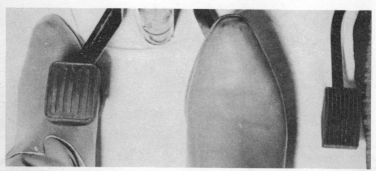

You decide to adjust speed, so you pivot the right foot on the heel and gently depress the centre pedal.

Once the road is all clear again, take your foot off the brake pedal and back to the accelerator.

This time you're in fourth gear, approaching a red traffic light and may have to stop.

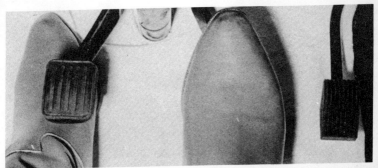

The right foot lifts off the accelerator and pivots on the heel to the brake. Having adjusted your speed with the footbrake, engage second gear. Then...

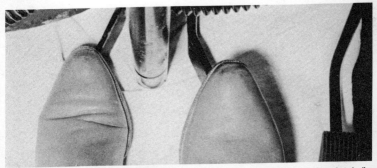

Apply the footbrake again (braking should be smooth), then depress the clutch (in practice just before the car stops) — to prevent the engine stalling.

Above and left: After the car has stopped, and not before, apply the handbrake by depressing the button and, keeping it depressed. Pulling the lever upwards as far as it will go. Then release the button. Again, remember that you mustn't release the brake if the car is on a slope. block or chock a wheel first!

You can now take your foot off the brake, but continue to keep the clutch down (see above). Move the gear lever from second to neutral. Now foot off clutch.

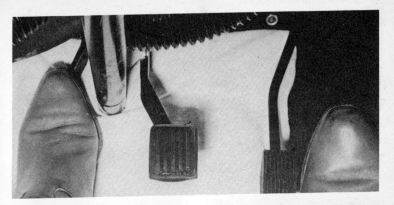

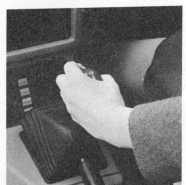

Above and below: The lights change, so down with the clutch and select first gear. Take your hand off the gear lever. Now push down the accelerator slightly and keep your right foot completely steady in that position.

At the same time gradually bring up the left foot.

Above and below: As the clutch pedal comes up, the engine note would change if the engine was running. So you would keep the clutch pedal held at this point. Then smoothly release the handbrake by depressing the button and, keeping it depressed, pushing the lever to the floor as far as it will go. Release button. Slowly lift clutch until fully raised and accelerate away.

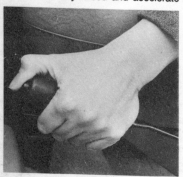

CHAPTER 5

STEERING GUIDELINES

Hold the wheel firmly but lightly in the 'ten-to-two' clock-face position (see below). Always use both hands unless changing gear, operating indicators, hand signalling, etc. When turning, never cross the hands.

STEERING GUIDELINES

Steering left

As the left hand pulls the wheel down from 10 to 7 o'clock the right hand slides from 2 to 5 o'clock. Simulate the correct movements by sliding your hands around the wheel, gripping slightly tighter with one hand so that you know which hand would be pulling on the wheel if you were moving.

The right hand then pushes the wheel from 5 to 2 o'clock as the left slides from 7 to 10 o'clock.

To straighten up, reverse these movements. As the right hand pulls the wheel down from 2 to 5 o'clock, the left hand slides from 10 to 7 o'clock... and so on.

STEERING GUIDELINES

Steering right

As the right hand pulls the wheel down from 2 to 5 o'clock, the left hand slides from 10 to 7 o'clock.

The left hand then pushes the wheel from 7 to 10 o'clock as the right slides from 5 to 2 o'clock.

To straighten up, reverse these movements. As the left hand pulls the wheel down from 10 to 7 o'clock, the right hand slides from 2 to 5 o'clock ... and so on.

Steering positions when reversing

When reversing, the same applies. At the moment of starting, the driver has to release the handbrake so the right hand holds the wheel at 2 o'clock. Without taking his eyes from his direction of travel, he has to return his left hand 'blind' to 10 o'clock. Practise this while looking over your left shoulder until you can consistently place the left hand on the wheel accurately.

Rating Test: 2

(Answers on page 126)

Tick the appropriate definitions:

1a Speed limit.
 b Clearway – no stopping.
 c Hospital.

2a No waiting at any time.
 b No loading at any time.
 c Waiting permitted for 20 minutes.

3a Do not enter junction unless your exit is clear.
 b Dangerous junctions — take care.
 c Special road surface safety experiment.

RATING TEST: 2

4a Right hand lane closed.
  b Two-way traffic ahead.
  c You may overtake in centre lane only.

5a No reversing.
  b No U turns.

6a Location of ungated level crossing.
  b No entry.
  c Motorway exit.

7 What does a single broken white line with long markings and short gaps on the road indicate?

  a Centre line.
  b No overtaking or crossing at any time.
  c Hazard warning line. Overtake or cross only if road ahead is well clear.

RATING TEST: 2

8 At what times should the horn not be used in a built-up area?

a 2000hrs-0200hrs.

b 2330hrs-0700hrs.

c 1700hrs-2330hrs.

9 At traffic lights when lights show red and amber, you must...

a Stop.

b Start to move off.

c Go if you are past the stop line.

10 What procedure should you adopt at a junction, entry to which has double broken white lines marked on the road?

a Stop.

b Proceed as you have precedence.

c You must slow down and be ready to stop to let traffic on the major road go by first.

Your score out of ten ....

(Mark the result on the graph on page 128)

CHAPTER 6

LOOK BACKWARDS FORWARDS

Eh, what's that again? Well, it would scarcely be conducive to safety if a driver had to look over his shoulder every few minutes to see what was happening behind. In no time at all, he would find the front of his car embedded in the

LOOK BACKWARDS FORWARDS

back of a vehicle in front ... or worse.

Which is why the law demands that all cars are fitted with a rear-view mirror.

It should always be referred to before signalling, starting, stopping, changing direction, and opening your door. So as we are about to simulate real life driving situations, get into the habit *now* of regularly glancing in it.

As will be stressed until the end of this chapter, it's mirror *first*, signal, and then whatever manoeuvre you're attempting.

Mirror

LOOK BACKWARDS FORWARDS

Signal

Manoeuvre

CHAPTER 7
STARTING AWAY PRACTICE

You are in the car and about to move away from stop. First check that the handbrake is on... and that the gearstick is in neutral.

STARTING AWAY PRACTICE

then assess the road conditions. If this was 'for real' you would be wearing your seatbelt.

Depress clutch (middle left) . . . select first gear (below left) . . . and depress button on handbrake ready for release (middle right).

Check the mirror (below right) . . .

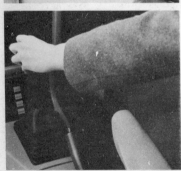

STARTING AWAY PRACTICE

signal right (top left)... push the accelerator down a trifle and keep your right foot completely steady in that position, at the same time slowly bringing up the left foot (middle). Pause slightly before lifting the clutch right up to simulate the car beginning to strain against the handbrake...

Glance over your right shoulder.

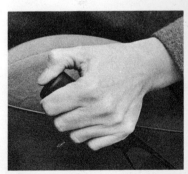

Release handbrake.

STARTING AWAY PRACTICE

Slowly raise clutch.

Steer slightly right.

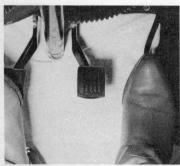

As the clutch is fully raised, begin to accelerate gradually. Straighten up steering.

Another mirror check.

CHAPTER 8

TURNING RIGHT

The turning right manoeuvre leads to more road accidents than any other single cause. So it's as well to really pay attention to this section. Mirror check, signal, position and speed during the approach are vital. So is keeping alert — you must watch for traffic on the road you are joining as well as the one you are leaving. If you are coming from a major road and are forced to wait at the intersection, where you stop is very important. Whatever you do, don't block the path of vehicles turning right from the road you're going to enter.

Be sure to signal in plenty of time when turning right.

You are in top gear and driving along a main road. Ahead there is a crossroads where you have precedence and where you want to turn right.

So glance in the mirror to check whether any vehicle is about to overtake you. There is another car some distance behind.

He is driving at the same speed, so put on the right indicator in good time, steer to just left of centre of the road and straighten up, indicator still on.

Check the mirror again.

Ensure the right indicator is still on.

Check your speed — apply footbrake.

Change down to third gear.

Change down to second gear.

You are now at the crossroads and going at a crawl — look ahead, right, left and right again.

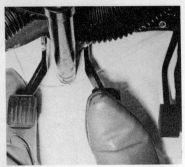

Cover your footbrake just in case...

Look ahead, right, left and right again. Glance in the mirror (someone might just attempt to overtake you)

... and, if all is clear, steer right.

After turning, straighten up. Check for pedestrians, and cancel signal.

You have joined a new road — so look in your mirror.

The road is clear and all is well, so accelerate away.

TURNING RIGHT

Turn right variations

There are a number of situations involving right turns — T junctions, traffic lights (conventional and filter), junctions controlled by policemen on point duty, intersections with traffic lane approaches... and U turns.

Simulate them all and practise all the variations: major road to minor, minor road to major — with some approaches guarded by Give Way signs and others by Stop signs.

Repeat each one several times until you know instinctively what action you must take. (For U turns, don't forget to glance over your right shoulder before turning.)

But remember those vital approach procedures: mirror, signal, manoeuvre, check speed, cover brake (stopping if necessary), and look!

CHAPTER 9

TURNING LEFT

You are in top gear and driving along the nearside of a main road. Ahead there is a minor road on the left and you wish to turn into it. Note the road sign (right).

So glance in the mirror to check behind.

There is another vehicle following you at the same speed so put on the left indicator in good time to give him advance warning of your intentions.

Check the mirror again.

Leave the indicator on.

Begin to check your speed – apply the footbrake.

Change down to third gear.

Change down to second gear.

Cover your footbrake . . . just in case. And check that there are no pedal or motor cyclists on your nearside. Then . . .

Begin to steer left. Turn corner.

Straighten up ... then cancel signal.

You have joined a new road – so look in your mirror.

All is well, so accelerate away.

TURNING LEFT

Turn left variations

There are as many varieties of left turns as there are of right, so refer back to Page 59. Apart from U turns, simulate them all, and repeat until you know instinctively what to do.

Meanwhile ... on to roundabouts. As with all junctions, the routine is: mirror, signal, manoeuvre, position, speed and ... look!

In the great majority of cases, the only vehicles that have precedence are those already going round.

The other rules depend on the exit you intend to take. If turning left or going straight ahead, use the left lane on approach, keep to the nearside as you go round and signal as you pass the exit before the one you are aiming for.

When turning right, enter in the right hand lane, and keep to the offside as you go round. Keep the right hand indicator on until you reach the exit before the one you want. Then put on your left hand indicator and switch lanes with care.

CHAPTER 10

IN REVERSE

The handbrake is on and the gear is in neutral. You are parked on the nearside of the road and about to reverse. So clutch down (right) . . .

Select reverse gear (below left).

Depress the button on the handbrake ready to release (below right).

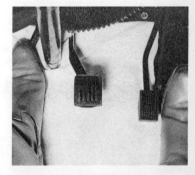

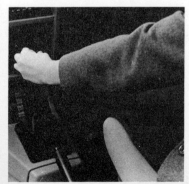

Check all around (above)... glance in the mirror (above right)... look over the left shoulder (below).

... Push the accelerator down slightly and keep your right foot completely steady in that position, at the same time slowly bringing up the left foot. Pause slightly before letting the clutch right up to simulate the car beginning to strain against the handbrake... release handbrake, and, when completely off, let go of the button... now you're moving, so it's both hands on the steering wheel.

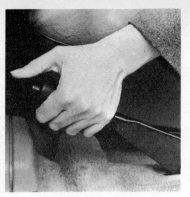

Practice parking

Another manoeuvre involving reversing is parking between two vehicles. Draw up alongside the leading car then, after taking the usual precautions, reverse and steer hard left almost immediately. Directly your front wing clears the lead car, take full lock right and straighten up.

The turn in the road by means of forward and reverse gears? Follow the routine for starting forwards, except that you take full lock right. Four feet (1.2m) from the opposite kerb, steer hard left, checking your speed with footbrake and clutch. Stop a foot (30 cm) from the gutter, apply handbrake and put the gear in neutral before lifting the clutch.

Check that your way is clear, keep left lock, reverse and, four feet (1.2m) from the other path, take right lock. Again check speed and stop about a foot (30 cm) clear, handbrake on and out of gear. Check the road and, keeping right lock, complete the manoeuvre.

Both these exercises can be simulated at home (a reverse turn into a side road – *never* back into a main one – is explained on Pages 83 and 84).

Final tip: whenever in reverse, go *very* slowly.

Rating Test: 3

(Answers on page 126)

1 What is the difference between 'Give Way' road markings and 'Stop' road markings?

   ....

   ....

2 There are three situations when you should test your footbrake. Name them.

   ....

   ....

   ....

3 Your car is parked on the nearside of the road, you have already started the engine, the handbrake is on, and you are about to set off. Write down the numerical sequence of the procedures that need to be taken (NB: One procedure is repeated more than once).
   a Mirror
   b Look over right shoulder
   c Clutch down
   d Gear
   e Signal
   f Clutch up, accelerator down, release handbrake.

RATING TEST: 3

4   The correct procedure when turning right at a roundabout is:
a   No signal necessary on approach; take either lane, signalling only at exit point.
b   Approach in left-hand lane, use left indicator before entering and then change to right-hand indicator. Then back to the left-hand indicator before the exit to be taken.
c   Approach in right-hand lane. Use right-hand indicator before entering and maintain this signal, changing to left-hand indicator at the exit before the one to be taken.

5   You may overtake a vehicle ahead of you on its left when . . .
a   It is in the outside lane of a motorway but going slower than you.
b   It is obvious that the vehicle is going straight on and is not likely to swing across to the left.
c   The vehicle ahead has signalled that it is turning right.

6   When would you use headlights in daytime?
a   Never; it is an offence under the Road Traffic Act.
b   As a sign to warn other vehicles you are overtaking.
c   In conditions of poor visibility.

7   As a general rule on a roundabout, who has precedence?
a   The vehicle approaching the roundabout.
b   The vehicle which is on the roundabout.

RATING TEST: 3

c   The vehicle which is approaching on a main A road.

8   A blue circular sign with a white 30 indicates:
a   Maximum speed.
b   Minimum speed.
c   Maximum speed for vehicles of more than three tons.

9   If you carry an advance warning triangle sign, what is the least distance it should be placed from the rear of the vehicle on roads other than motorways?
a   50 yds (46m).
b   100 yds (91m).
c   25 yds (23m).

10   A green arrow at traffic lights means:
a   Filter as indicated regardless of other lights.
b   Filter only when green is shown on main traffic lights.
c   Give way to approaching traffic only.

Your score out of 10 ....
(Mark the result on the graph on Page 128)

GETTING READY

The moment you've been waiting for...

Up to now you've been sitting in a stationary car, engine off, not moving or going anywhere. All the situations on the road, the way the car reacts to the pedals, have all been inside your imagination. Hopefully you have started to develop a feel for the car's controls, and a sense of the rhythm and the sequences of driving a car on the road.

Now it's time to try out some of those movements, but this time with the engine running. We don't want to venture on to the public road yet. All we want to do now is to get you accustomed to the sound of the engine when it has been started, controlling its speed with the accelerator, and the feel of the clutch and the gear lever when the power is actually flowing through the transmission of the car.

Now where you do this is important. Obviously we don't want you trying out clutch control in a restricted drive in front of a garage! If things go wrong and you have a moment of panic, you'll be through those garage doors before you can think. We want a bit of space, but off the public highway. You obviously will not be able

GETTING READY

to take the car to such a location yourself; someone will have to drive you there and remain with you while you practise. Your local road safety officer at the town hall may be able to help with a location. Some councils even have a special private road system for learners.

Bear in mind the fact that you will be in control of a motor vehicle. If you are going to use an empty car park, or anywhere the public has access to, make sure all the paperwork is correct. You will already have a provisional driving licence. If the vehicle is yours, you will have already insured it, and you will have made sure the person with you has a full licence and is covered to drive your car. If it's not your car, you will

The correct driving position. The elbows and knees are half-bent, enabling the hands to grasp the steering wheel effortlessly in the 'ten-to-two' position and the feet to reach the pedals easily.

GETTING READY

have made sure that the insurance covers you.

On the way to and from the practice site, don't just stare out of the window looking at the scenery. Fill the time usefully by studying every move the driver makes, and equate each one with the exercises you have done to date. There's always more to learn.

If you don't want to take instruction from the driver when you reach the practice area you don't need to. Work together from this book. Get into the driving seat, and make sure you are as comfortable as you were when you were carrying out your simulated driving earlier in the stationary car.

Next, ask your companion to explain all the controls, dials and gadgets. Learn their whereabouts and what they do off by heart. In fact, don't proceed any further until you can automatically locate them all without the slightest hesitation.

Remember the correct gearstick grasp; the ball of your hand over the knob at the top and a gentle grip of the stick immediately below with thumb and forefinger.

Cockpit drill next . . . and this is something you *must* get off by heart. Before starting the engine, you have got to check that the gear is in neutral, the handbrake is on, all doors are closed, the driving seat is in the right position, the mirrors are clean and properly adjusted, the lights, indicators and horn are working, and that the seatbelts fit and are fastened.

Only when you have completed this drill can you start up. And the checks aren't over yet. Before moving off, look at the instruments to ensure that you have petrol, sufficient oil and that the red ignition light goes out when the accelerator is pushed down. Make sure, too, that the wipers and windscreen washers are

GETTING READY

working. Two final points while on the move: test the footbrake immediately after setting off (ensure it is safe to do so), and remember to return the choke control (if the car has a manual choke) if you have used it.

Now repeat the gearchanging exercise — this time with the engine switched on. *But keep the handbrake on and don't take your feet off the clutch until you return the gear lever to neutral.*

Next, get the feel of the accelerator — with the engine on. Ensure that the gear lever is in neutral, depress the accelerator pedal *gradually*, and ease off gently when the engine begins to 'scream'.

Keeping the engine ticking over — *and with the handbrake still on* — combine the last two exercises. Depress the clutch and select first gear. Now slightly push down the accelerator and hold it rock-steady. bring up the clutch very slowly until you feel the car just begin to strain against the handbrake. Repeat this exercise several times. Then try the same procedure in reverse gear.

Rating Test: 4

(Answers on page 126)

1   To manoeuvre a car out of a rear wheel skid, the driver should first:
a   Apply the brakes.
b   Steer straight ahead.
c   Steer in the direction of the skid.
d   Steer against the direction of the skid.
e   Remove foot from accelerator.

2   When may you enter a bus lane?
a   Never.
b   At the times laid down on the road signs.
c   When traffic outside the lane is at a standstill.
d   At any time.
e   At any time, providing there are no buses using it.

3   When may you park your vehicle on the right hand side of the road at night?
a   Until midnight, if the street lights are on.
b   Never.
c   At any time, providing your parking lights are on.
d   Only in a one-way street.

RATING TEST: 4

4 You are parked on the nearside of the road. The engine is running, the gear lever in neutral and the handbrake on. You are all set to make a turn in the road by means of forward and reverse gears. Write down the numerical sequence of the procedures to be taken (NB: Some procedures are of course repeated more than once).

a Look all around.
b Accelerator down and clutch up.
c Reverse gear.
d Clutch down.
e First gear.
f Mirror.
g Steer right to full lock.
h Handbrake off.
j Look over left shoulder.
k Clutch down, brake gently, and stop.
m Steer left to full lock.
n Handbrake on and select neutral.
p Steer straight.

5 You are approaching a crossroads where a policeman is on point duty. You wish to indicate to him that you are going straight ahead. What signal do you use?

6 What does this illuminated sign mean?
a Move into lane shown.
b Leave motorway at next exit.
c Inside lane closed ahead.

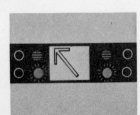

(Yellow lights on blue background and black arrow outline on white)

RATING TEST: 4

7 What do these road signs mean?
a Count-down markers on motorways only.
b Count-down markers on motorways and primary routes.
c Count-down markers to a level crossing.

 (White diagonals on blue)

8 What does this sign mean?
a No through road.
b All lanes closed on the motorway.
c T junction ahead.

(Red-topped T on blue)

9 What does this road sign mean?
a Sharp deviation.
b Count down markers.
c Turn left: one way only.

10 Amber at traffic lights means . . .
a Prepare to move off.
b Go.
c Stop.

Your score out of 10 ....
(Mark the result on the graph on page 128)

CHAPTER 12

ON THE MOVE...

Now you are ready to drive the car round the practice area. This part of the syllabus is the final stage of your preparation for on-the-road tuition from a professional instructor.

You will learn how to judge the length and width of the car, how to park, reverse, turn left and right, make turns in the road by means of forward and reverse gears — in fact just about every manoeuvre you will ever be called upon to make.

Practise over and over again until you perfect each one — and, as your competence grows, so will your confidence.

First things first, though: let's get the car actually on the move this time. Don't forget to fasten your seat belt, and to make sure your passenger has fastened his, *BEFORE* you start the engine and move off.

Depress the clutch, select first gear, and grasp the handbrake ready for release. Take a little accelerator, *hold it dead steady*, then slowly bring up the clutch. As you feel the car start to pull, release the handbrake.

No more accelerator yet — let it stay where it is, but

ON THE MOVE

smoothly and gently up with the clutch. If the car begins to 'Kangaroo', just push the clutch back down, check that the handbrake is completely off, and put a shade more pressure on the accelerator. Then, gradually, up with the clutch again. Keep plugging away until you achieve total smoothness.

Once you've got the hang of this and can change gear smoothly, pull in, put the car in neutral, pull on the handbrake, and try reversing.

While mastering clutch, gear and accelerator co-ordination, pay attention to steering technique by simply putting into practice what you learnt at home. Directly you feel in control, concentrate on those vital mirror-signal-manoeuvre routines each time you start, stop or turn.

Resist the urge to look at your feet and the gear lever, keep your eyes glued to where you're going, and glance frequently in the mirror. Talking of mirrors, check that they have been adjusted properly.

Make sure that the driving seat is positioned correctly then, first, adjust the interior mirror. It should give you an uninterrupted view through the rear window without any need to move your head.

Both you and your passenger should fasten your belts.

Adjust the interior mirror to give an uninterrupted view behind.

Adjust the wing mirror to give you the maximum view at the side of the car.

Wing-mounted mirrors will need to be adjusted by a friend.

Most cars have door mirrors which you can adjust. If your car has the old fashioned wing mirrors, get your companion to set them for you.

The most important of these is the one on the offside. In it you should be able to see the front of an overtaking vehicle as it draws abreast with the rear of your car (see top left picture opposite).

With the nearside mirror, you should be able to see the road and pavement alongside the car and beyond (see top right picture opposite).

Offside mirror view.

Nearside mirror view.

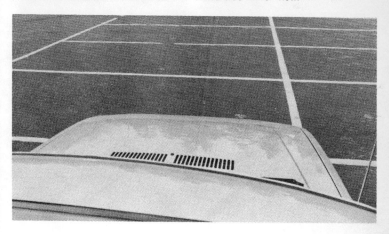

Kerbdrill — car style

Now you're all set. So go through your cockpit drill... and start the engine.

Find a line on your driving area, a continuous break in the road surface, a kerb, or a painted line. Or use strips of wood.

Drive round and draw up as close as you can to it without touching it (the view through the windscreen is shown in the picture above). Continue until you can

ON THE MOVE

effortlessly stop the car parallel with the kerb and about six inches (15 cm) away from it.

Next, try it in reverse. Stop the car in such a position that you can see the whole of the kerb through the rear window over your left shoulder. Once again, aim to pull up about six inches (15 cm) away from it.

In both instances, get your companion to tell you how you are doing (above).

The long and short of it

Now place the cartons you have brought along with you about 7 ft 6 in (2.3 m) apart. Drive slowly through the gap forwards and backwards without touching the boxes (see pictures opposite). As your confidence and speed increase, steadily reduce the distance between the cartons until they are only six inches (15 cm) away from the side of the car.

Having perfected that, approach the gate from different angles.

Do this consistently without hitting the boxes and you're well on your way to judging widths from the driving seat.

Back and so forth?

Now, length judgement. Place the cartons about 25ft (7.6 m) in front of and behind the car. drive slowly forwards and backwards, stopping as close to the walls as you can without touching them (see picture above and top picture on page 84). Keep practising until you can consistently pull up within six inches (15 cm) of the boxes.

With the boxes, form a street corner, you are about

ON THE MOVE

to try some reverse turns (see picture on page 85).

Stop the car in such a position at the nearside kerb that you can see the corner through the rear window.

After checking all around, look over your left shoulder, steer straight back ... then left to full lock at the bend. Straighten up so the car is parallel with and six inches (15 cm) away from the kerb.

There are several exercises you can try with the 'props' — left and right turns at minor and major roads, T junctions, crossroads and roundabouts, for instance.

You can also simulate parking between two vehicles at the roadside, and practise turns in the road by means of forward and reverse gears.

For the first, place the boxes 20 feet (6 m) apart and 4 feet (1.2 m) from the roadside. For the second, lay out two parallel lines 25 feet (7.6 m) apart.

To refresh your memory on the methods of these manoeuvres, refer to page 67.

Remember: practice makes perfect. But there's a big, big difference between driving off the highway where there are no pedestrians and other vehicles to contend with, and driving on the road.

That is why, from here on, professional instruction is essential.

Rating Test: 5

(Answers on pages 126 and 127)

1. You are on your driving test and the examiner tells you to stop beyond a minor road on your nearside. He then asks you to reverse into it. As you go by, you look into the minor road to ensure that the manoeuvre can be carried out safely — then pull up ready to reverse. The engine is running, the gear in neutral and the handbrake on. Write down the numerical sequence of the procedures to be taken.
   a Accelerator down gently, clutch up.
   b Look over left shoulder.
   c Look into the junction.
   d Clutch down.

RATING TEST: 5

- e Reverse gear.
- f Steer straight.
- g Handbrake off.
- h Cover footbrake.
- j Straighten up quickly.
- k Out of gear, foot off clutch.
- m Handbrake on.
- n Clutch down, brake gently, and stop.
- p Steer hard left.
- q Steer right.

2 You are starting away and the car begins to jerk uncomfortably. What are the two likeliest causes?

a ....

b ....

3 Some of the above road signs are merely acting as warnings; others are giving you an order. By each

RATING TEST: 5

one, mark 'W' for warning or 'O' for order where appropriate.

4 You are behind the wheel and about to set off. But before starting the engine you must go through your cockpit drill. What is the full sequence?

....

....

....

5 After starting up, there are seven more checks to be made. What are they?

....

....

....

6 If the red ignition light on the dashboard comes on while you are driving, what could it mean? Tick the appropriate faults.
   a The engine has stalled.
   b The dynamo or alternator is giving trouble.
   c The fan belt has snapped.
   d The engine has overheated.
   e You are running low on oil.
   f The battery is flat.
   g There is a short-circuit.

RATING TEST: 5

7 You are stopped by a red traffic light at a crossroads. After a wait of several minutes, it is obvious that the lights have stuck. Can you legally proceed past the red light?
 a Yes.
 b No.

8 If you do not have exterior mirrors, can you legally drive on the highway if your boot lid is up and totally obstructing the view through the rear window?
 a Yes.
 b No.

9 If your car is classified as a four-seater, do you break the law if the vehicle is occupied by five adults?
 a Yes.
 b No.

10 You are about to drive off in darkness. You are satisfied that your side, tail, brake, head and indicator lights are functioning. What other lamp have you omitted from your check list?

....

Your score out of 10 ....
(Mark the result on the graph on Page 128)

CHAPTER 13

OBSERVATION

'I just didn't see it...'

Ask any policeman for the most common lament they hear when interviewing drivers after accidents and it will almost certainly be this one.

The sad truth is that many drivers simply haven't acquired the skill of reading the road — the art of recognising certain clues and building up a mental picture of 'hidden' hazards ahead.

Just how much is missed is shown up in independent tests when candidates have been asked to recollect a road sign they have just passed. In too many cases, they have been unable to do so and signs are the most obvious clue of all!

Road observation, which demands the utmost concentration and knowing instinctively where the eyes should look, is a craft that has to be worked at. And one of the best ways to learn is to keep up a silent running commentary of what you're seeing, its significance, and how you will react to it.

Look at the pictures of real life road situations on page 90. In the first one, we are on an urban dual

OBSERVATION

OBSERVATION

carriageway road, and we've just been overtaken by the van. What is he going to do now? He may be turning right across the centre reservation at the junction ahead. If he slows down and we come up on his inside, will our way be clear ahead? Is there anything approaching the junction on our left, possibly hidden behind the bushes, a pedestrian, cyclist or motor bike? We need to use the mirror, and check our speed. Be prepared to slow down or brake.

In the second picture, we are on a fast country road, coming into a village. We will be slowing for the 40 mph (64 kph) limit sign anyway, but will the pick-up truck on the right hand side of the road try to nip out in front of us, anticipating that we will be slowing down. Mirror, check speed, be prepared to slow or brake. What are the hazards ahead, beyond the road junction? More crossroads and junctions. Start planning route, course and action now so that when you arrive you are in the right gear, travelling at the right speed to cope with all eventualities.

In the third picture, we are in an urban street, so drive with great caution. Look for tell tale signs of a driver about to pull out — smoke from exhaust, brake lamps on, driver in seat and passenger walking over to get in. This is a prime area for playing children. If a child runs out from behind the parked cars on the left, *can you stop?* Leave as much room as you can on the nearside and look out for doors being flung open. Watch the whole street, not just the nearside. There's more danger and accident potential here than in the thick of the motorway in the rush hour!

Rating Test: 6

(Answers on page 127)

1. You are in top gear and approaching a T junction. You are on the minor road. Your entry is guarded by a Give Way sign and you wish to turn right. Write down the numerical sequences of the procedures to be taken (NB: Two procedures are repeated more than once).

   a Move to just left of the centre of the road.
   b Signal.
   c Slow right down and select appropriate gear.
   d Straighten up.
   e Mirror.
   f Cancel signal.
   g Steer right if safe to do so.
   h Look right, left and right again.
   j Cover footbrake.

2. What does this road sign mean?
   a Uneven road surface.
   b Muddy banks.
   c Falling or fallen rocks.

3. You are approaching a crossroads and the traffic lights are green. You intend to turn left. According to the Highway Code, you may carry out the manoeuvre if the way is clear. What else does it say?

   ....

RATING TEST: 6

4 You have a frozen windscreen and haven't time to remove all the frost. Can you legally drive the car if you have cleared an area a foot (30 cm) square directly in front of you?
  a Yes.
  b No.

5 What is the maximum blood-alcohol level permitted when driving?

.... milligrams to 100 millilitres.

6 What is the overall stopping distance on a dry road with a vehicle in first class condition when you are travelling at 30 mph (48 kph)?
  a 25 ft (7.6 m)
  b 50 ft (15 m).
  c 75 ft (23 m).
  d 100 ft (30 m).

7 Most car skids are due to . . .
  a Over-inflated tyres.
  b Under-inflated tyres.
  c Uneven tyre pressures.
  d Slippery surfaces.
  e Speed too fast for road conditions.

8 You are travelling at 50 mph when a blow-out occurs in the nearside front tyre. After lifting off the accelerator, should the driver . . .
  a Apply the brakes?
  b Steer to the left?

RATING TEST: 6

    c Steer straight ahead?
    d Steer to the right?

9 Are good driving skills more important than good driving attitudes?
    a Yes.
    b No.

10 When driving at 60 mph behind another vehicle what is the minimum safe following distance?
    a 30 yds (27 m).
    b 40 yds (36 m).
    c 50 yds (46 m).
    d 60 yds (55 m).

Your score out of 10 ....
(Mark the result on the graph on Page 128)

CHOOSING AN INSTRUCTOR

Can you imagine the confusion of trying to learn a language with one teacher telling you to pronounce the words one way, and another telling you to do it completely differently? In circumstances like that, the chances of becoming proficient and passing an exam would obviously be much reduced. Yet that is how a great number of people learn to drive.

For they get friends or relatives to show them the ropes — very often between lessons from a motoring school. The trouble is that they run a very real risk of being taught the wrong things by drivers who have not been trained to teach.

Bad habits drivers may have developed since casting aside their L-plates are all too easily passed on. The mixture of wrong advice from this source with correct methods taught by professional instructors can only result in confusion. Which, in turn, can severely shorten the odds of passing the test.

Has Dad, Uncle Bill, or lifelong pal, Fred, for example, heard of DL68? Probably not. yet it's the L-driver's 'bible' — the free booklet issued by the

95

CHOOSING AN INSTRUCTOR

government explaining what is required to prove competency to a driving examiner, and how it is achieved. A professional instructor, on the other hand, knows it off by heart, upside-down and backwards!

There is only one way to learn to drive — properly! That means professional training. By all means, take

CHOOSING AN INSTRUCTOR

advantage of a relative or friend's offer to take you out for practice, but make sure that's as far as it goes. Practice, yes, as much as you can get, teaching, no.

If you have the option, restrict your practice to one type of car — preferably the same make and model supplied by the driving school. Attempting to drive very different vehicles from the one you are being taught in, could prove confusing and unnerving and lead to pressures you can well do without on test day.

How do you choose an instructor — more important, the right instructor? One way is to ask people you know. Remembering their own experiences when they were L-drivers, they might be able to recommend somebody personally.

Failing that, have a word with an official of your local driving instructors' association. Such organisations exist in many areas — and the public library should be able to tell you how to get in touch. If there isn't one where you live, ask the advice of the road safety unit at your nearest council offices.

When you have made your choice, do ask to see the instructor's credentials. By law, professional instructors have to be on the government's 'approved list'.

There are two types of identity card and both are pictured on page 96. The one at the top is issued to fully-qualified teachers; the other to trainees who haven't yet passed all the official examinations.

It is important that you should take this precaution. For there are a number of so-called instructors operating illegally. Their prices may be attractive . . . but, in the absence of qualifications, their standards must be in doubt. In the long run, it will pay you to choose wisely.

HOW TO PASS THE DRIVING TEST

Many people get very worried and concerned in the days leading up to the driving test. In many cases, they are so nervous and tense, they find it impossible to drive to the best of their ability, and they end up losing confidence and putting up a poor performance which inevitably ends in failure. Possibly the best advice is to give yourself so much experience of driving, in a variety of situations, that the test itself holds no fears for you at all. If you are at likely to forget how to do a turn in the road; if you have trouble reversing round a corner; if you tend to muff the hill start — practise the manoeuvres over and over again, until you have complete confidence in your own ability. Once the mechanics of driving have been mastered, the psychological problem of the test will be enormously reduced.

After all, the examiner is not trying to trick you or catch you out. He simply wants to know if you can handle a car competently and confidently, and the primary and overriding aim is to ensure that you pose no threat, through incompetence, to the other people on the road.

HOW TO PASS THE DRIVING TEST

However, don't fall into the trap of thinking that the test is a mere formality before you breeze on to the highway in your first Maserati. Over-confidence can lead to lack of attention and concentration, both of which will cost you your pass.

Let's just run through some of the standard manoeuvres the examiner will ask you to perform, and look at some of the ways you can use the car to make things a bit easier.

When you get into the car with the examiner, put on your seatbelt, and wait for him to do the same. Make sure the handbrake is tight on, and waggle the gear stick around quite obviously to ensure it is not in gear. Set the interior mirror so that you have a clear view through the rear window. Start up, and allow the engine to idle for a few seconds while you get yourself settled, calm and collected and you know just what you want to do. The examiner will give you instructions about where he wants you to go. Don't make the mistake of following the test route your instructor used to take you over during your lessons.

Before moving off, remember to wind down your window, even if it's raining and bitterly cold, and take a good look behind. Don't start to move until you are looking to the front again, and as you accelerate away from the kerb, check the mirror. Don't pull out in front of anyone, don't force any oncoming cyclists, cars or other road users to brake or take any avoiding action. If the road is not clear, wait. The examiner won't mind.

Once under way, maintain a reasonable pace. Remember the speed limit and observe it, and come down well under that speed in any sort of hazardous situation. Don't crawl around on your test, and do try to give the examiner a smooth ride at all times. Remember to keep three or four feet from the edge of the kerb to maintain a safety line.

Don't dread meeting lorries, cyclists, pedestrians or horse-riders — welcome them! They all provide you with an opportunity to show the examiner how well you can handle the situation. Keep a sharp look-out. If you can anticipate something happening on the road in front of you, and take appropriate action, you are providing the examiner with concrete proof of your competence. Watch for car doors being flung open from parked cars ahead. Look out for pedestrians and especially children.

The examiner will want you to run through a series of manoeuvres to see how well you handle the car. He will want you to do a hill start. When he asks you to halt on a steep section of road, remember, *mirror, signal, manoeuvre* and *brake*. Practise stopping safely and correctly at the side of the road. If you find it difficult to bring the car to a halt with the tyres close to the kerb, you can use a little trick. When you are sitting in the driving seat, with the car parked about six inches (15 cm) from the kerb, note exactly where on the bonnet, or the wing the kerb comes into view at the front of the car. mark that spot with a bit of coloured tape. Then when you slow and pull into the kerb, you will know that if you line up your mark with the kerb, you will always be stopping in exactly the right place.

When you make the hill start, accelerate strongly, but don't race the engine. Don't release the handbrake as soon as the clutch comes up. You must not run backwards one inch. Leave the handbrake fully on, with your hand round the handle, and your thumb pressing in the release button. Lift the clutch pedal until the clutch starts to bite, and watch the nose of the car closely. As soon as you see it dip, indicating that the power is reaching the wheels, release the handbrake, lift the clutch all the way, increase the power, and away you go. Do it smoothly, slowly and gently. Practise, practise, practise, on the steepest hills you can find. Do it again and again until you think you can do it in your sleep!

You will also probably be told to stop by the kerb just past the mouth of a junction, and then to reverse back round the corner keeping an even distance from the kerb. The trick is to do things very smoothly, and to watch your accelerator control. Learners tend to pump the accelerator pedal up and down when they reverse, making the car lurch. The lurch throws the driver forward so that he puts more weight on the throttle, and in no time a see-sawing movement results. Practise driving smoothly backwards, changing speed gradually and gently.

Some people find it difficult to reverse at a constant distance from the kerb. They weave in and out, often finishing up at an angle to the kerb, and well away from it. Again, a visual aid can help a lot. Sit in the driving seat with the car parked near the kerb, and look out of the back window. Get someone to place an ordinary steel pin in the lower edge of the window rubber at the spot where the kerb can be seen emerging from behind the car. Make sure your head is in the position it would be if you were reversing, and that your hands are on the wheel. Now, when reversing, simply line up the pin head to the kerb, and you will always be the same

HOW TO PASS THE DRIVING TEST

distance away from it. Remember to use small movements of the wheel to change direction when reversing. Heavy swings don't do any good at all.

The examiner will ask you to stop at the nearside kerb. Then he will ask you to turn the vehicle in the road, in as few turns as possible, using forward and reverse gears. The road the examiner chooses may well be relatively narrow and have quite a pronounced camber, so the car will tend to roll up against the kerb — if it is allowed to. The secret is to treat the turn in the road as separate hill starts, except that at least one of them will be in reverse. When you are asked to carry out the manoeuvre, you will be expected to turn immediately, turning the wheel quickly, passing it from hand to hand — *without crossing your hands* — to get the car round on to full lock as quickly as possible. Don't forget to check your mirror, and wait for all traffic to go before starting.

With the front of the car swinging towards the opposite kerb, bring the speed down, turn the wheel back to the other lock in the last few feet, and stop short of the kerb. Apply the handbrake.

When you change into reverse, increase the revs gently, ease up the clutch, and wait until you feel the tail of the car squat slightly as the power reaches the axles, before you let off the brake. Check again that the way is clear. Get on to full lock as fast as you can when reversing, but watch your speed, and just before you bring the car to a halt at the opposite kerb, quickly take off the lock. Apply the handbrake again, and don't let the car roll into the kerb.

Again, increase the revs, ease up the clutch and wait till the nose dips before releasing the handbrake. Keep checking for safety, and don't accelerate before you have the car straight and under control.

The examiner will also ask you to make an emergency stop. He will move his clipboard smartly to the windscreen, at the same time saying 'Stop'. You have to remember that he is simulating a situation in which a child dashes out from the kerb, right under your wheels. You have to stop the car immediately, under control, so that the examiner can assess your speed of reaction.

Don't be frightened of the car. Stamp on the brake pedal hard, then press down the clutch just before the car stops. You must do this in a split second. If you bring the car to a halt slowly and smoothly, the examiner will fail you, because you have not proved you can stop your vehicle quickly in an emergency. Again, practise on your own. If you do practise on the public road, make sure there is no-one behind you before you do your stop and don't spring an emergency stop on your passenger as a surprise, if he's unbraced he will be thrown into his seat belt with an unpleasant jar.

Finally, read the Highway Code again and again until you know it inside out. Many people have put up a superb driving performance and failed on the Highway Code questions. Don't let yourself become one of them. Test yourself, and get others to test you until you know it backwards. Not only will it help you pass your driving test, it will also make you a much better driver!

BASIC MAINTENANCE

Each time before you set out...

...It is vital that you give your car a good look over.

Sad to relate, few motorists give their cars the briefest glance before starting a journey. Also on the findings from the MoT test and independent inspections, they're not too careful about having their vehicles serviced, either. Small wonder that a recent study has shown that one in eight car accidents are directly or indirectly caused by defects.

Yet the law's demands at the back of the Highway Code clearly require that, 'the condition of your vehicle is such that no danger is likely to be caused to yourself or others'.

So get into the habit of taking simple precautions before setting out — like looking for signs of oil, petrol, brake fluid and water leaks on ground over which your car has been standing.

And ensure that you have ready to hand everything you might need: replacement light bulbs, an unopened

BASIC MAINTENANCE

canister of the correct grade of hydraulic fluid, at least a quart of the right engine oil, spare anti-freeze (again the correct sort) an accurate tyre pressure gauge, battery topping-up mixture and a tin of grease.

Remember: safety consciousness starts *before* you get into the car.

Look at the tyres

Three simple checks on each tyre (and don't forget the spare) should be carried out to ensure that the pressures are as stated in the vehicle's handbook (do

Keep the engine compartment clean and wash off any excess oil or grease with a proprietary engine cleaner. This will enable you to keep clean when doing maintenance, and allow you to spot any leaks.

BASIC MAINTENANCE

this with the pressure gauge, (see picture above, top), the tread depth is at least 2mm all round (for safety), and that the tread surface and both walls are sound (see picture above, bottom left).

Now the lights

'To see and be seen' is a most important safety code. It is therefore essential that *all* your lights work—they

BASIC MAINTENANCE

must, at all times, by law in any case. See that they do (picture opposite, bottom right) by switching them on and walking round the car (don't forget the indicators, headlamp dipping system and rear number plate light). To check the brake lights, try catching their reflection in a window.

Keep those windows clean

If the windows aren't clear, you can't see *properly*. It's as simple as that. So clean them — inside and out (above). In winter, make sure that the windscreen and rear window are completely frost, snow or ice-free. Also give the lights and mirrors a wipe over.

Then up with the bonnet

Having done that, raise the bonnet for more essential checks.

BASIC MAINTENANCE

 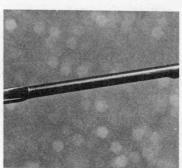

Check the oil and water

Locate the dipstick which should be withdrawn and wiped clean (above left). Replace and remove once more. If the oil trace is below the minimum mark on the dipstick, top up (above right).

Now, check the coolant level in the radiator — or equalising tank if it's a sealed system. Top up as necessary. If the system is filled with anti-freeze mixture, replenish it with the correct grade and ratio of antifreeze to water.

Check the brake fluid

With the aid of the car handbook, locate the brake fluid reservoir and check the level. If it is between the minimum and maximum marks, top up. Then operate the brake pedal several times and re-check the level. If it has fallen get the brake system seen to immediately. If the level is below minimum, don't use the vehicle until the system has been inspected for leaks, topped up and bled. Also check the clutch fluid reservoir level.

BASIC MAINTENANCE

The screen washer flask too

If your car is fitted with a windscreen washer system, it must work at all times and the reservoir (above left) always be topped up. A useful tip here: add a few drops of methylated spirit until the water begins to cloud.

Once a month...

With a wheelbrace, check that the wheel nuts are tight (above right). (When changing a wheel, ensure that the rounded ends of the nuts go on first.) Also inspect the condition of all water, fuel, brake and vacuum servo hoses.

... and every three months

1 Lubricate the jack lightly, greasing or oiling the threads.
2 Look under the floor covering and the *car's underside* for corrosion so that early action can be taken before

BASIC MAINTENANCE

trouble becomes more serious and widespread.

If using a jack, also place supports under the raised tyres to avoid accidents.

Keep a note of your mileage. As a general rule, contact breaker points should be changed every 5,000 miles (8,000 km) and sparking plugs serviced every 5,000 miles. If your car won't start on a damp morning it is probably moisture or condensation on plug leads or inside the distribution cap. Wipe all the leads, and the inside of the cap (see below), and use one of the anti-damp sprays from your local DIY shop. Refer to the car's handbook for directions and also for items like air filter replacements. (NB: Some air filter intakes have to be positioned differently for summer and winter motoring.)

Finally . . . enrol at a local evening institute for a car maintenance course. If you cannot spare the time for DIY servicing, ensure that the car goes to a garage at the recommended service intervals specified in the handbook.

Rating Test: 7

(Answers on page 127)

1 You have just made your daily checks on the car before setting off. Which ones have been omitted from the following list: Side lights, dipped and main beam headlamps, indicators, water, oil, petrol, tyres' condition and pressures, and cleaning of windows and lamps.

....

2 Tick the definitions which apply to the zig-zag markings on the road at pedestrian crossings...
a Pedestrians should not cross the road at these points.
b Motorists have the right of way if the vehicle is in the zig-zag zone before the appearance of a pedestrian who wishes to use the crossing.
c No overtaking.
A further legal requirement has been omitted. Which one?

....

3 The police have the right to ask you to produce your driving licence, insurance certificate and MoT test certificate. Do traffic wardens have the same powers?
a Yes.
b No.

RATING TEST: 7

4   What has been omitted from this legal requirement: 'Pedestrians, pedal cycles, motor cycles under 50 cc capacity, invalid carriages not exceeding 5 cwt unladen weight, certain slow moving vehicles carrying oversized loads (except by special permission), agricultural vehicles and animals must not use motorways.'

....

5   Which of these describes the road sign denoting two-way traffic ahead?
    a Black upward and downward pointing arrows on a white background in a red circle.
    b A black upside-down Y on a white background in a red triangle.
    c Black upward and downward pointing arrows on a white background in a red triangle.
    d A black upside-down Y on a white background in a red circle.

6   How often, according to the manufacturers, should brake hydraulic fluid be replaced?
    a Once very 3 years.
    b Once every 2 years.
    c Every year.

7   What should be the minimum thickness of ...
    a A bonded brake lining?
    b A disc brake pad?

RATING TEST: 7

8 When taking off the wheels and hubs to inspect the brake linings or pads, what else should you especially be looking for?

....

9 Your battery is in good order and fully charged, yet your lights and radio go off suddenly. What are the first two things you should check?

....

10 Alternate flashing red lights on any road sign mean what?

....

Your score out of 10 ....
(Mark the result on the graph on Page 128)

Rating Test: 8

(Answers on page 127)

1 How often, according to the manufacturers, should a car's brake system be completely overhauled?
  a Every two years or every 25,000 miles (40,000 km).
  b Every three years or every 40,000 miles (65,000 km).
  c Every four years or every 50,000 miles (80,000 km).

RATING TEST: 8

2 You are approaching a crossroads and wish to turn left, but have inadvertently driven into a 'straight ahead, turn-right' lane. Can you legally turn left, providing you clearly signal your intention, if there is traffic behind you, and moving vehicles passing you on the nearside?
a Yes.
b No.

3 You have switched off the engine, undone your seatbelt and are about to get out of the car. You are parked on the nearside of the road. What procedure should you follow?

....

....

4 Which definition describes the road sign denoting that you must give priority to vehicles from the opposite direction?
a A large black arrow pointing upwards and a large black arrow pointing downwards on a white background with a red diagonal line through the upward pointing arrow (bound by a red circle).
b A small red arrow pointing upwards and a large black arrow pointing downwards on a white background (bound by a red triangle).
c A large black arrow pointing upwards and a large black arrow pointing downwards on a white background with a red diagonal line through the upward pointing arrow (bound by a red triangle).

RATING TEST: 8

  d A small red arrow pointing upwards and a large black arrow pointing downwards on a white background (bound by a red circle).

5 What is meant by the 'totting up' system?
  a The drink-drive law.
  b The right of the police to remove your vehicle if parked illegally.
  c The accumulation of driving licence endorsements.

6 The red light at a Pelican crossing has changed to a flashing amber signal. What does this mean?

  ....

7 Where would you expect to find alternately flashing red lights? And what action would you be required to take?

  ....

8 You have just put brand-new tyres on your car. Over how many miles should they be run in?
  a 50 miles (80 km).
  b 100 miles (160 km).
  c 200 miles (320 km).
  And what should be the maximum speed during this period?
  d 40 mph (64 kph).
  e 45 mph (72 kph).
  f 50 mph (80 kph).

RATING TEST: 8

9 Officially, flashing headlamps may be used:
   a To indicate that you give way.
   b To indicate that you do not intend to give way.
   c In place of your horn.

10 Study this picture. You are driving down the road — so there are four things in particular that you should be watching for. What are they? And what two actions should you be taking?

....

....

Your score out of 10 ....
(Mark the result on the graph on page 128)

CHAPTER 17
FIRST AID

'Well, frankly, I didn't want to get involved...'

'Of course I *would* have stopped and helped, but what was the use? There was nothing I could have done...'

Excuses, excuses — and very sadly the all too familiar reason why the lives of dozens of road accident victims are lost each year when they could easily be saved.

All because so few motorists know even the basic rudiments of First Aid. Motorists in Britain, that is...

For years, more enlightened European nations like West Germany have insisted that their drivers be 'genned up' on the subject. Indeed, you have to know First Aid out there by law if you're a motorist. Moreover, it is an offence not to carry an emergency kit in your vehicle.

Wherever he may be, it is *every* good driver's duty to be versed in the necessary knowledge. And it doesn't mean giving up a string of evenings to learn, either. The St. John Ambulance Brigade run a special two-hour

FIRST AID

rapid course especially for motorists. Nor is it expensive. Your local branch will be only too pleased to hear from you.

All in all, a small price to pay for 'know-how' which, should you ever be called on to put into practice, could end up with YOU saving somebody's life.

You would be surprised how the simplest action can save a life . . .

Take an accident which happened near Bournemouth recently. A farm tractor, emerging from a field, was involved in a collision — and the impact caused two cars to crash head-on.

Seconds afterwards, two vehicles passed the scene, but failed to stop — their drivers plainly intent on 'not getting involved'.

Fortunately it was a case of 'third time lucky.' For the third car to arrive at the spot did stop. And the driver — a woman — had recently attended that special St. John Ambulance Brigade First Aid course.

She immediately took charge, sending the shocked but uninjured tractor driver off to telephone for an ambulance, then taking a close look at the casualties in the nearest of the two cars. Happily, none was seriously

First aid kit

hurt and, after assuring them that an ambulance was on the way, she dashed across to see the other motorist involved.

As it turned out, he wasn't alright. He wasn't badly injured . . . but his face was blue. The force of the crash had thrown him on to his back, causing his tongue to block his airway.

Quickly she pressed his head backwards. And with his neck extended and his lower jaw pushed upwards, his tongue moved forward, opening the air passage.

That is all she did. And in a trice, the man was breathing normally again.

There would have been a quite different outcome if that woman had been a member of the vast 'don't know' brigade. By the time the ambulance had arrived, he would have been dead.

If you're one of the uninitiated, just remember what that woman did. Treat it, if you will, as your first lesson in First Aid.

Despite what you might think, there's no great mystery about First Aid. Although it is based on the principles of practical medicine and surgery, it is largely a matter of commonsense. Unfortunately when the chips are down and things start to happen, the first course of action that flashes into an untrained person's mind is not always the right one.

It's a sobering thought that a wrong decision can finish off what an accident may already have half-done — and that's kill the victim. Instinctively know what you're about, on the other hand, and you'll automatically come up with the *right* answer.

So let's take a look at the simple measures that anyone can grasp — methods which, if carried out

FIRST AID

properly, quickly and gently, can be life-saving and may prevent the necessity for more complicated treatment later which could come *too* late.

Right then, you're driving along a main road at night when you suddenly spot it: a car embedded in the roadside fence. You at once pull in to the left and, stopping only to collect your First Aid kit and a torch dash to the accident scene.

The front nearside door of the battered vehicle is already open. You switch off the ignition and, in the glow of the courtesy light, you set to work, assessing the degree of the occupant's injuries.

The two in the front seat look in a bad way. They are both unconscious. A third person — in the back — is conscious. You talk to him while attending to the two in front and learn that he has a nasty cut just above the knee. You tell him to elevate his leg and give him a tissue to press on the wound.

Fortunately, another car has pulled up and you call urgently for assistance. you have already sized up the situation and it is clear that you will have to deal with the front seat passenger first. Unlike the driver, he doesn't appear to be breathing.

So, with the aid of the newcomer, gently ease him out into the open, lay him down and cover him with your overcoat.

Now there's no time to lose. Kneeling down behind his head, you support the nape of his neck with one hand and tilt his head backwards with the other, thus ensuring a clear airway. This done, you press the angles of his jaw forward to improve that airway.

If he doesn't start breathing now, his condition is serious. He doesn't . . .

FIRST AID

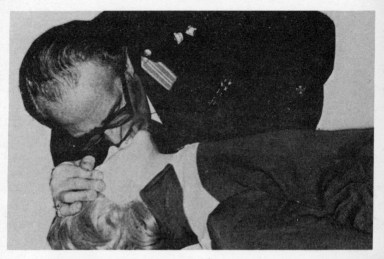

You shift position to the left of the injured man's head and, still supporting the nape of his neck, tilt his head back in readiness to prepare him for mouth-to-mouth resuscitation (see picture above). While you're doing this, you send your helper back to the road to get someone to 'phone for an ambulance.

Keeping the victim's head extended, you open your mouth wide and take a deep breath. You pinch his nostrils with your fingers then seal your lips around his mouth. You blow, hoping to see his chest rise. But there seems to be a blockage. You feel around his mouth and at the back of his throat and remove the obstruction.

You again open your mouth wide, take a deep breath and, pinching his nostrils, seal your lips around his mouth. you blow and, this time, his chest rises. You remove your mouth and watch his chest deflate. You give three more inflations in double quick time with the aim of saturating the man's blood with oxygen. After that, you slow down to around 10 inflations a minute. If a victim doesn't start breathing on his own, you *must* continue with resuscitation until medical aid arrives. But after a couple of minutes, with great relief, you see that the man is breathing on his own.

Now you put him in the recovery position — in other words on his side with his upper leg and arm extended at right angles to the body. This will prevent the man from drowning in his own gastric secretions.

By this time, your helper has returned and you both lift out the unconscious driver. You have checked that he is breathing properly in between treating the other injured man and now, covering him with your assistant's coat, you place him in the recovery position.

You ask your assistant to stay with the driver and front seat passenger and to advise immediately if there is any change in their condition. Now, you attend to the man in the back, and the first thing is to reassure him that an ambulance is on the way. While doing this, you take a look at his cut leg. It is bleeding profusely.

You gently lay him on his back on the rear seat, keeping his injured leg elevated. Then, with the pair of scissors in the emergency kit, cut away his trouser leg. With a sterile dressing squeeze the two edges of the wound together and pack the dressing into the depth of the wound until it projects above the surface of the leg. Cover with padding, and bandage firmly.

Ten minutes later, the ambulance and a police car arrive. All three, you hear next day, are 'comfortable'.

The question is: Would they have survived if, like the majority of motorists, you hadn't known where to start? The answer to that is up to you . . .

Rating Test: 9

(Answers on page 128)

1 If you come across a casualty who is not breathing, there are three essential steps you must take without delay. What are they?

....

....

....

2 If these do not work, it will be necessary to administer mouth-to-mouth resuscitation. How is this done?

....

....

3 There are two permanently fixed 'Stop' traffic signs. Which of these definitions are correct?
a The word 'stop' in red on a white background in an upside-down red triangle.
b The word 'stop' in black on a white background in an upside-down red triangle.
c The word 'stop' in red on a white background in a red circle.
d The word 'stop' in black on a white background in a red circle.

RATING TEST: 9

e The word 'stop' in red on a white background in an upside-down red triangle bounded by a red circle.

f The word 'stop' in black on a white background in an upside-down red triangle bounded by a red circle.

g An eight-sided, white-edged sign with the word 'stop' in white on a red background.

h An eight-sided, red-edged sign with the word 'stop' in red on a white background.

4 Double yellow lines along the edge of the carriageway indicate:
  a No stopping.
  b No loading or unloading.
  c No waiting except for loading or unloading.

5 If you are given a third licence endorsement within two years, what is the most likely penalty a Court will impose?

....

6 Can you legally take your car on the road after the previous tax disc has expired, assuming you have made a renewal application by post?
  a Yes.
  b No.

7 If you come across a casualty who is unconcious, you should put him in the recovery position. What is it?

....

RATING TEST: 9

8 Three yellow lines painted on the kerb indicate:
   a No loading at stated times during the working day.
   b No loading at any time.
   c No loading at odd hours stated on a plate.

9 You come across a casualty who is bleeding badly. How should you deal with the wound?

....

....

10 You are driving in the offside lane of a dual carriageway section when the car in front indicates that it is turning right. There is no traffic in the nearside lane. Write down the numerical sequence of the procedures you will need to take. (NB: One procedure is repeated more than once.)
   a Signal left.
   b Cancel signal.
   c Mirror.
   d Steer left to pass the car on the nearside when safe to do so (if you think this is legally permitted).

Your score out of 10 ....
(Mark the result on the graph on page 128)

ANSWERS

Rating Test One
1(c); 2(a); 3(a); 4(b); 5(a); 6(b); 7(b); 8(c); 9(b); 10(a).

Rating Test Two
1(b); 2(b); 3(a); 4(a); 5(b); 6(a); 7(c); 8(b); 9(a); 10(c).

Rating Test Three
1 'Give Way' (single or double white broken lines across the carriageway), 'Stop' (single or double white solid lines across the carriageway);
2 Immediately after starting away, traversing a ford, going through a deep puddle; 3 a(3,7), b(5), c(1), d(2), e(4), f(6); 4(c); 5(c); 6(c); 7(b); 8(b); 9(a); 10(a).

Rating Test Four
1(e); 2(b); 3(d); 4 a(3,12,22), b(4,14,23), c(11), d(1,10,20), e(2,21), f(27), g(6,17,25), h(5,15,24), j(13), k(8,18), m(7,16), n(9,19), p(26); 5. Press the right hand flat against the windscreen with the fingers pointing upwards; 6(a); 7(a); 8(a); 9(a); 10(c).

Rating Test Five
1 a(4), b(3), c(8), d(1), e(2), F(6), g(5), h(9), j(11), k(14), m(13), n(12), p(7), q(10); 2 The clutch has been released too quickly, the handbrake is on; 3. Left-to-right — Top row; O-W-O-W, centre: W-O-O-O, Bottom: O-W-W-O.
4. Check that the gear is in neutral, the handbrake is on, all doors are closed, the driving seat is in the right position, the mirrors are clean and properly adjusted, the lights, indicators and horn are working, and that the seatbelts fit and are fastened.
5. Check the instruments to ensure that you have petrol, sufficient oil, and that the red ignition light goes out when the accelerator is depressed. Make sure the wipers and windscreen washer are working. Test footbrake after starting off (when safe to do so), and remember to return the choke control if you have used it. 6 (a,b,c,g);

ANSWERS

7(b); 8(b); 9(b) providing the extra passenger doesn't obstruct the car's controls or the rear view mirror; 10 Rear number plate.

Rating Test Six

1 a(3), b(2), c(5), d(11), e(1,4,9,13), f(12), g(10), h(6,8), i(7); 2(c); 3. Take special care and give way to pedestrians who are crossing; 4(b); 5 Eighty; 6(c); 7(e); 8(c) the first step is to keep the car on the road; 9(b); 10(d).

Rating Test Seven

1 The rear registration number plate lamp, brake fluid level, screen washer tank; 2 (a,c), No stopping except to give precedence to a pedestrian or in circumstances beyond your control, or when it is necessary to avoid an accident; 3(a); 4 Learner drivers; 5(c); 6(c); 7 a(1/16in); b (⅛in); 8 Scoring of drums (if appropriate), contamination of braking surfaces by oil leaking brake fluid, corrosion and dust, leaking brake cylinders, pipes and connections, the condition of the discs and self-adjusting mechanisms (if appropriate), and correct adjustment of other parts and components; 9 Fuses or battery connections; 10 Stop.

Rating Test Eight

1 (b); 2 (b) - a good driver will proceed on his wrongly-chosen course until he can safely regain his route, making a detour if necessary; 3 Ensure that the gear is in neutral and the handbrake is on, that everything is switched off (unless parking lights are required) and cigarettes have been extinguished. Then check the mirror or look over your right shoulder before opening the door. Finally, see that windows, doors and the boot are properly closed and locked (but don't forget to leave some ventilation if you are leaving a pet in the car); 4 (d); 5(c) 6 Give way to any pedestrians on the crossing but otherwise you can proceed. 7 Level crossings and urban motorways — and you must stop; 8 (b,f); 9(c); 10 Watch for animals, pedestrians (especially children) emerging from between parked vehicles, car doors opening and cars pulling out. So check your speed and, in the prevailing conditions pictured, steer a centre line.

ANSWERS/TEST GRAPH

Rating Test Nine

1 Lay the casualty on his back on a firm surface, check that he has a clear airway and, if not, improve it.
2 Open mouth wide, take a deep breath, pinch the casualty's nostrils with the fingers, seal you lips around his mouth. Blow until the casualty's chest rises, and remove mouth. After the casualty's chest deflates, repeat sequence — the first four inflations should be given rapidly; 3 (f), (g); 4 (c); 5 Disqualification; 6(b); 7 Place the casualty on his side with his upper leg and arm extended at right angles to the body; 8(b); 9 Place as sterile dressing on the injury and squeeze the edges of the wound together. pack the wound with dressing until it projects above the injury. Cover with padding and bandage firmly. 10a(2) — there may be no traffic in the nearside lane, but there may be someone behind you, b(4), C(1,5), d(3).

TEST GRAPH

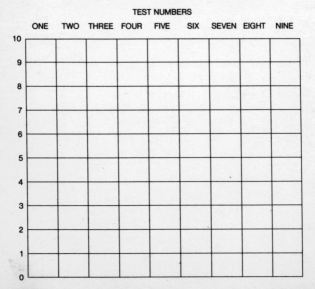